From Clay to Bricks

From Clay to Bricks

Ali Mitgutsch

 Carolrhoda Books, Inc., Minneapolis

First published in the United States of America 1981 by
Carolrhoda Books, Inc. All English language rights reserved.

Original edition © 1975 by Sellier Verlag GmbH, Eching bei München,
West Germany, under the title VOM LEHM ZUM ZIEGEL.
Revised English text © 1981 by Carolrhoda Books, Inc.
Illustrations © 1975 by Sellier Verlag GmbH.

Manufactured in the United States of America

LIBRARY OF CONGRESS CATALOGING IN PUBLICATION DATA

Mitgutsch, Ali.
  From clay to bricks.

    (A Carolrhoda start to finish book)
    First published under title: Vom Lehm zum Ziegel.
    SUMMARY: Highlights the step-by-step process of
digging clay and forming it into bricks.

    1. Brickmaking—Juvenile literature. [1. Brickmaking]
I. Title.

  TP827.M5413   1981   666'.737   80-29567
  ISBN 0-87614-149-1

        2 3 4 5 6 7 8 9 10 86 85 84 83 82

From Clay to Bricks

The ground we walk on is made up of many different things.

Rock, sand, and clay are all found in the ground.

Clay is really tiny pieces of soil that become

soft and stretchy when mixed with water.

Clay is used to make bricks.

First it must be dug out of the ground.

Then the clay is taken to a brick factory.

At the factory it is put into a machine called a **dry pan**

where it is crushed with heavy rollers.

When the clay comes out of the dry pan,

any small rocks that are still in it are taken out.

Then water is added.

The clay turns into a thick, stretchy paste.

Next the clay is sent through a press
that squeezes out all of the air.
When it comes out of this machine,
the clay looks like a long brown sausage.

Now the clay is cut into pieces
with a tightly stretched wire.
The pieces of clay look like bricks now.
But they are still soft and moist.

The bricks are put into a large drying room to dry out.

The air in this room is heated as high as 300°F (150°C).

When the bricks come out of the drying room,

they are hard and dry.

But if they get wet, they will become soft again.

So they must be burned to make them waterproof.

This is called **firing**.

The bricks are stacked in a huge oven called a **kiln**.

Then coal is shoveled between the stacks.
When the coal is burned, the temperature in the kiln
gets very hot (1600°-2000°F, 871°-1100°C).
This makes the bricks waterproof and very strong.

At last the bricks are ready to be used.
They are taken to a bricklayer
who is building a house.
The bricks will make a beautiful,
strong house for someone to live in.

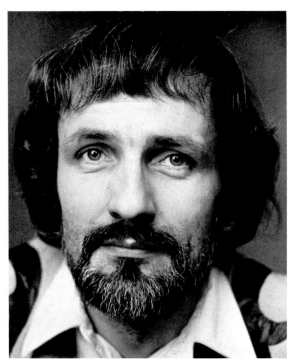

**Ali
Mitgutsch**

ALI MITGUTSCH is one of Germany's best-known children's book illustrators. He is a devoted world traveler, and many of his book ideas have taken shape during his travels. Perhaps this is why they have such international appeal. Mr. Mitgutsch's books have been published in 22 countries and are enjoyed by thousands of readers around the world.

Ali Mitgutsch lives with his wife and three children in Schwabing, the artists' quarter in Munich. The Mitgutsch family also enjoys spending time on their farm in the Bavarian countryside.

THE CAROLRHODA

START

From Beet to Sugar

From Blossom to Honey

From Cacao Bean to Chocolate

From Cement to Bridge

From Clay to Bricks

From Cotton to Pants

From Cow to Shoe

From Dinosaurs to Fossils

From Egg to Bird

From Egg to Butterfly

From Fruit to Jam

From Grain to Bread

From Grass to Butter

From Ice to Rain

From Milk to Ice Cream

From Oil to Gasoline

From Ore to Spoon

From Sand to Glass

From Seed to Pear

From Sheep to Scarf

From Tree to Table

TO FINISH

BOOKS